P9-CBR-508

LTX
7-12-0
Var

Connected Mathematics™

Variables and Patterns

Introducing Algebra

Student Edition

DISCARDED

Glenda Lappan
James T. Fey
William M. Fitzgerald
Susan N. Friel
Elizabeth Difanis Phillips

LAMAR UNIVERSITY LIBRARY

Developed at Michigan State University

DALE SEYMOUR PUBLICATIONS®
MENLO PARK, CALIFORNIA

Connected Mathematics™ was developed at Michigan State University with the support of National Science Foundation Grant No. MDR 9150217.

This project was supported, in part,
by the
National Science Foundation
Opinions expressed are those of the authors
and not necessarily those of the Foundation

The Michigan State University authors and administration have agreed that all MSU royalties arising from this publication will be devoted to purposes supported by the Department of Mathematics and the MSU Mathematics Education Enrichment Fund.

This book is published by Dale Seymour Publications®, an imprint of Addison Wesley Longman, Inc.

Dale Seymour Publications
2725 Sand Hill Road
Menlo Park, CA 94025
Customer Service: 800 872-1100

Managing Editor: Catherine Anderson
Project Editor: Stacey Miceli
Revision Editor: James P. McAuliffe
Book Editor: Mali Apple
Production/Manufacturing Director: Janet Yearian
Production/Manufacturing Coordinators: Claire Flaherty, Alan Noyes
Design Manager: John F. Kelly
Photo Editor: Roberta Spieckerman
Design: PCI, San Antonio, TX
Composition: London Road Design, Palo Alto, CA
Electronic Prepress Revision: A. W. Kingston Publishing Services, Chandler, AZ
Illustrations: Pauline Phung, Margaret Copeland, Ray Godfrey
Cover: Ray Godfrey

Photo Acknowledgements: 16 (Chicago) © Jean-Claude Lejeune/Stock, Boston; 16 (Melbourne) © Joe Carini/The Image Works; 18 © Rick Smolan/Stock, Boston; 22 © Fredrik Bodin/Stock, Boston; 23 © William Johnson/Stock, Boston; 24 © Larry Mulvehill/The Image Works; 32 © Duomo, Inc.; 36 © Joe McBride/Tony Stone Images; 42 © Addison Geary/Stock, Boston; 51 © E. Strenk/Superstock, Inc.; 57 © Topham/The Image Works

Copyright © 1998 by Michigan State University, Glenda Lappan, James T. Fey, William M. Fitzgerald, Susan N. Friel, and Elizabeth D. Phillips. All rights reserved. No part of this publication may be reproduced, stored in a retrieval system, or transmitted, in any form or by any means, electronic, mechanical, photcopying, recording, or otherwise, without prior written permission of the publisher. Printed in the United States of America.

RAGBRAI is a registered trademark of the Des Moines Register and Tribune Co.
Turtle Math is a registered trademark of LCSI.

DALE SEYMOUR PUBLICATIONS®

Order number 45840
ISBN 1-57232-645-X

5 6 7 8 9 10-BA-01-00 99

The Connected Mathematics Project Staff

Project Directors

James T. Fey
University of Maryland

William M. Fitzgerald
Michigan State University

Susan N. Friel
University of North Carolina at Chapel Hill

Glenda Lappan
Michigan State University

Elizabeth Difanis Phillips
Michigan State University

Project Manager

Kathy Burgis
Michigan State University

Technical Coordinator

Judith Martus Miller
Michigan State University

Curriculum Development Consultants

David Ben-Chaim
Weizmann Institute

Alex Friedlander
Weizmann Institute

Eleanor Geiger
University of Maryland

Jane Mitchell
University of North Carolina at Chapel Hill

Anthony D. Rickard
Alma College

Collaborating Teachers/Writers

Mary K. Bouck
Portland, Michigan

Jacqueline Stewart
Okemos, Michigan

Graduate Assistants

Scott J. Baldridge
Michigan State University

Angie S. Eshelman
Michigan State University

M. Faaiz Gierdien
Michigan State University

Jane M. Keiser
Indiana University

Angela S. Krebs
Michigan State University

James M. Larson
Michigan State University

Ronald Preston
Indiana University

Tat Ming Sze
Michigan State University

Sarah Theule-Lubienski
Michigan State University

Jeffrey J. Wanko
Michigan State University

Evaluation Team

Mark Hoover
Michigan State University

Diane V. Lambdin
Indiana University

Sandra K. Wilcox
Michigan State University

Judith S. Zawojewski
National-Louis University

Teacher/Assessment Team

Kathy Booth
Waverly, Michigan

Anita Clark
Marshall, Michigan

Theodore Gardella
Bloomfield Hills, Michigan

Yvonne Grant
Portland, Michigan

Julie Faulkner
Traverse City, Michigan

Linda R. Lobue
Vista, California

Suzanne McGrath
Chula Vista, California

Nancy McIntyre
Troy, Michigan

Mary Beth Schmitt
Traverse City, Michigan

Linda Walker
Tallahassee, Florida

Software Developer

Richard Burgis
East Lansing, Michigan

Development Center Directors

Nicholas Branca
San Diego State University

Dianne Briars
Pittsburgh Public Schools

Frances R. Curcio
New York University

Perry Lanier
Michigan State University

J. Michael Shaughnessy
Portland State University

Charles Vonder Embse
Central Michigan University

Special thanks to the students and teachers at these pilot schools!

Baker Demonstration School
Evanston, Illinois

Bertha Vos Elementary School
Traverse City, Michigan

Blair Elementary School
Traverse City, Michigan

Bloomfield Hills Middle School
Bloomfield Hills, Michigan

Brownell Elementary School
Flint, Michigan

Catlin Gabel School
Portland, Oregon

Cherry Knoll Elementary School
Traverse City, Michigan

Cobb Middle School
Tallahassee, Florida

Courtade Elementary School
Traverse City, Michigan

Duke School for Children
Durham, North Carolina

DeVeaux Junior High School
Toledo, Ohio

East Junior High School
Traverse City, Michigan

Eastern Elementary School
Traverse City, Michigan

Eastlake Elementary School
Chula Vista, California

Eastwood Elementary School
Sturgis, Michigan

Elizabeth City Middle School
Elizabeth City, North Carolina

Franklinton Elementary School
Franklinton, North Carolina

Frick International Studies Academy
Pittsburgh, Pennsylvania

Gundry Elementary School
Flint, Michigan

Hawkins Elementary School
Toledo, Ohio

Hilltop Middle School
Chula Vista, California

Holmes Middle School
Flint, Michigan

Interlochen Elementary School
Traverse City, Michigan

Los Altos Elementary School
San Diego, California

Louis Armstrong Middle School
East Elmhurst, New York

McTigue Junior High School
Toledo, Ohio

National City Middle School
National City, California

Norris Elementary School
Traverse City, Michigan

Northeast Middle School
Minneapolis, Minnesota

Oak Park Elementary School
Traverse City, Michigan

Old Mission Elementary School
Traverse City, Michigan

Old Orchard Elementary School
Toledo, Ohio

Portland Middle School
Portland, Michigan

Reizenstein Middle School
Pittsburgh, Pennsylvania

Sabin Elementary School
Traverse City, Michigan

Shepherd Middle School
Shepherd, Michigan

Sturgis Middle School
Sturgis, Michigan

Terrell Lane Middle School
Louisburg, North Carolina

Tierra del Sol Middle School
Lakeside, California

Traverse Heights Elementary School
Traverse City, Michigan

University Preparatory Academy
Seattle, Washington

Washington Middle School
Vista, California

Waverly East Intermediate School
Lansing, Michigan

Waverly Middle School
Lansing, Michigan

West Junior High School
Traverse City, Michigan

Willow Hill Elementary School
Traverse City, Michigan

Contents

Variables and Patterns

When Ben started to play the electric guitar, his skill increased quite rapidly. But over time, as he continued to practice, he seemed to improve more slowly. Sketch a graph to show how Ben's guitar-playing skill progressed over time from when he first started playing. What variables might affect the rate at which his playing improved?

List three things about yourself and the world that change. Explain how some of these changes might be related to each other. For example, your height changes as your age changes. How would you represent these changes in a table or graph?

Compare the rental fees of two companies. East Coast Trucks charges $4.25 for each mile a truck is driven. Philadelphia Truck Rental charges $200 for a week, or any part of a week, and an additional $2.00 for each mile a truck is driven. Which of these two companies offers a better rate? If you were renting a truck, what variables might influence your decision?

All around you, things are changing. Temperatures and tides rise and fall, seasons and moon phases change, prices and the popularity of television shows vary. You are changing too. Your height, weight, hair length, and your moods and tastes are changing. Sometimes there is a relationship in the way two things are changing. For example, the number of hours of daylight changes as the seasons change, and the prices of goods change as consumer demand changes.

In mathematics, quantities that change are called *variables*. In this unit, you will explore variables and the relationships between them. You can show these relationships using things like tables, graphs, and equations. These tools are at the heart of the branch of mathematics called *algebra*.

As you work through the investigations in this unit, you will encounter problems like those on the opposite page.

Mathematical Highlights

In this unit you will begin to study algebra. You will learn some useful mathematical methods for studying patterns of change in the world.

- After you conduct an experiment involving jumping jacks, making a table and a coordinate graph of the results introduces you to two important ways to display data.

- Analyzing information given in tables, graphs, and written notes helps you discover the advantages and disadvantages of each of these forms of representation.

- As you examine the expenses and costs involved in running a small business, you see how tables and graphs can be important tools for making predictions and decisions.

- Writing general rules, or equations, to describe patterns of change lets you compute values of one variable for any value of the other variable.

- Making graphs or tables for equations provides a way to represent the patterns and compare the effects of making changes.

- If you use a graphing calculator, you can make tables and graphs quickly.

Using a Calculator

In this unit, you will be able to use your calculator to build tables of data that you intend to graph. As you work in the Connected Mathematics units, you may decide whether using a calculator will help you solve a problem.

Variables and Coordinate Graphs

The bicycle was invented in 1791. Today there are over 100 million bicycles in the United States. People of all ages use bicycles for transportation and sport. Many people spend their vacations taking organized bicycle tours.

Did you know?

RAGBRAI—which stands for Register's Annual Great Bicycle Ride Across Iowa—is a week-long cycling tour across the state of Iowa. It has been held every summer since 1973. Over 7000 riders dip their back bicycle wheels into the Missouri River along Iowa's western border, spend seven days biking through Iowa's countryside and towns (following a different route every year), and end the event by dipping their front bicycle wheels into the Mississippi River on the state's eastern border.

1.1 Preparing for a Bicycle Tour

The popularity of bicycle tours gave five college students—Sidney, Celia, Liz, Malcolm, and Theo—an idea for a summer business. They would operate bicycle tours for school and family groups. They chose a route from Philadelphia, Pennsylvania, to Williamsburg, Virginia, including a long stretch along the ocean beaches of New Jersey, Delaware, and Maryland. They decided to name their business Ocean and History Bike Tours.

While planning their bike tour, the five friends had to determine how far the touring group would be able to ride each day. To figure this out, they took test rides around their hometowns.

Think about this!

- How far do you think you could ride in a day?
- How do you think the speed of your ride would change during the course of the day?
- What conditions would affect the speed and distance you could ride?

To answer the questions above, you would need to take a test ride yourself. Although you can't ride your bike around the classroom, you can perform a simple experiment involving jumping jacks. This experiment should give you some idea of the patterns commonly seen in tests of endurance.

Problem 1.1

This experiment requires four people:
- a jumper (to do jumping jacks)
- a timer (to keep track of the time)
- a counter (to count jumping jacks)
- a recorder (to write down the number of jumping jacks)

As a group, decide who will do each task.

Prepare a table for recording the total number of jumping jacks after every 10 seconds, up to a total time of 2 minutes (120 seconds).

Time (seconds)	0	10	20	30	40	50	60	70	...
Total number of jumping jacks									

Here's how to do the experiment: When the timer says "go," the jumper begins doing jumping jacks. The counter counts the jumping jacks out loud. Every 10 seconds, the timer says "time" and the recorder records the total number of jumping jacks the jumper has done so far. Repeat the experiment four times so that everyone has a turn at each of the four tasks.

Problem 1.1 Follow-Up

Use your table of jumping jack data to answer these questions:

1. How did your jumping jack rate (the number of jumping jacks per second) change as time passed? How is this shown in your table?

2. What might this pattern suggest about how bike-riding speed would change over a day's time on the bicycle tour?

1.2 Making Graphs

In the jumping jack experiment, the number of jumping jacks and the time are variables. A **variable** is a quantity that changes or *varies*. You recorded your data for the variables in a table. Another way to display your data is in a coordinate graph. A **coordinate graph** is a way to show the relationship between two variables.

There are four steps to follow when you make a coordinate graph.

Step 1 *Select two variables.*

For example, for the experiment in Problem 1.1, the two variables are *time* and *number of jumping jacks*.

Step 2 *Select an axis to represent each variable.*

If time is one of the variables, you should usually put it on the x-axis (the horizontal axis). This helps you see the "story" that occurs over time as you read the graph from left to right. So, in a graph of the jumping jack data, time would go on the x-axis, and the number of jumping jacks would go on the y-axis (the vertical axis).

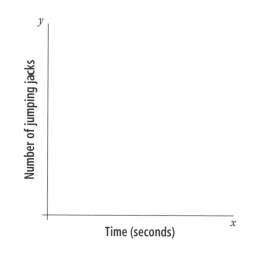

In many cases, you can determine which variable to assign to which axis by thinking about how the two variables are related. Does one variable *depend* on the other? If so, put the **dependent variable** on the *y*-axis and the **independent variable** on the *x*-axis. The number of jumping jacks depends on time. So, put number of jumping jacks (the dependent variable) on the *y*-axis and time (the independent variable) on the *x*-axis. You may have encountered the terms *dependent variable* and *independent variable* while doing experiments in your science classes.

Step 3 *Select a scale for each axis.*
For each axis, you need to determine the largest and smallest values you want to show on your graph and how you want to space the scale marks.

In the jumping jack experiment, the values for time are between 0 and 120 seconds, so in a graph of this data, you could label the *x*-axis (time) from 0 to 120. Since you collected data every 10 seconds, you could place marks at 10-second intervals.

The scale you use on the *y*-axis (number of jumping jacks) depends on the number of jumping jacks you did. For example, if you did 97 jumping jacks, you could label your scale from 0 to 100. Since it would be messy to put a mark for every jumping jack, you could put a mark for every 10 jumping jacks.

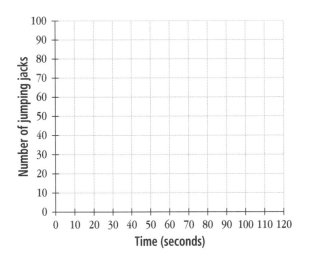

Step 4 *Plot the data points.*

For example, suppose that at 60 seconds, you had done 65 jumping jacks. To plot this information, start at 60 on the *x*-axis (time) and follow a line straight up. On the *y*-axis (number of jumping jacks), start at 65 and follow a line straight across. Make a point where the two lines intersect. This point indicates that in 60 seconds, you did 65 jumping jacks.

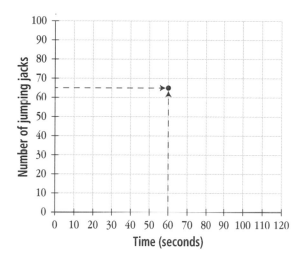

Problem 1.2

A. Make a graph of your jumping jack data.

B. What does your graph show about jumping jack rate as time passes? (Another way to say this is, What does your graph show about the *relationship* between the number of jumping jacks and time?)

■ Problem 1.2 Follow-Up

Is the relationship you found between the number of jumping jacks and time easier to see in the table or the graph? Explain your answer.

As you work on these ACE questions, use your calculator whenever you need it.

Applications

1. The convenience store across the street from Metropolis School has been keeping track of their popcorn sales. The table below shows the total number of bags sold beginning at 6:00 A.M. on a particular day.

  a. Make a coordinate graph of these data. Which variable did you put on the *x*-axis? Why?

  b. Describe how the number of bags of popcorn sold changed during the day. Explain why these changes may have occurred.

Time	Total bags sold
6:00 A.M.	0
7:00 A.M.	3
8:00 A.M.	15
9:00 A.M.	20
10:00 A.M.	26
11:00 A.M.	30
noon	45
1:00 P.M.	58
2:00 P.M.	58
3:00 P.M.	62
4:00 P.M.	74
5:00 P.M.	83
6:00 P.M.	88
7:00 P.M.	92

2. The graph below shows the numbers of cans of soft drink purchased each hour from a school's vending machine in one day (6 means the time from 5:00 to 6:00, 7 represents the time from 6:00 to 7:00, and so on).

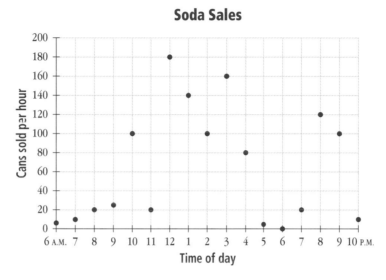

Soda Sales

a. The graph shows the relationship between two variables. What are the variables?

b. Describe how the number of cans sold changed during the day. Give an explanation for why these changes might have occured.

3. Below is a chart of the water depth in a harbor during a typical 24-hour day. The water level rises and falls with the tides.

Hours since midnight	0	1	2	3	4	5	6	7	8	9	10	11	12
Depth (meters)	10.1	10.6	11.5	13.2	14.5	15.5	16.2	15.4	14.6	12.9	11.4	10.3	10.0

Hours since midnight	13	14	15	16	17	18	19	20	21	22	23	24
Depth (meters)	10.4	11.4	13.1	14.5	15.4	16.0	15.6	14.3	13.0	11.6	10.7	10.2

a. When is the water deepest? What is the depth at that time?

b. When is the water shallowest? What is the depth at that time?

c. During what time interval does the water depth change most rapidly?

d. Make a coordinate graph of the data. Describe the overall pattern you see.

e. How did you determine what scale to use? Do you think everyone in your class used the same scale?

Connections

4. The mayor of Huntsville and her advisory board were trying to persuade a company to build a factory in the town. They told the company's owner that the population of Huntsville was growing very fast and would provide the factory with an abundant supply of skilled labor. A local environmental group protested, saying this company had a long history of air and water pollution. They tried to persuade the factory owner that the population was not increasing as fast as the mayor's group had indicated. The company hired their own investigator to research the situation. When the three parties met, each party presented a graph. The graphs are shown below and on the next page.

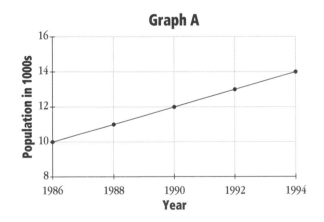

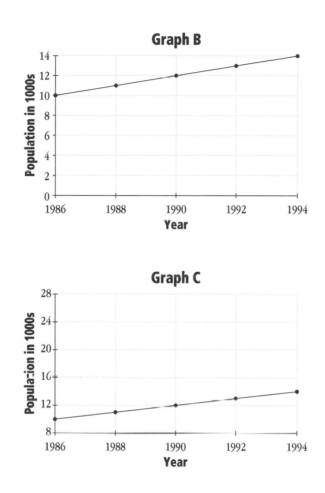

a. Which graph do you think was presented by the mayor? The environmentalists? The company's investigator? Explain your reasoning.

b. Is it possible that all the graphs correctly represent the population growth in Huntsville? Why or why not?

c. Describe the relationship between time and population as shown in the graphs.

5. After doing the jumping jack experiment, Andrea and Ken compared their graphs. Because his points were higher, Ken said he did more jumping jacks than Andrea in the 120 seconds. Do you agree? Why or why not?

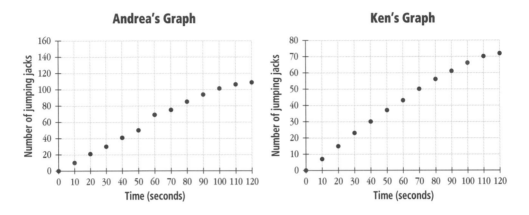

Andrea's Graph

Ken's Graph

6. The operators of Ocean and History Bike Tours wanted to compare their plans with other bicycle tour companies. The bike tour they were planning would take five days, and they wondered if this might be too long or too short for people. Malcolm called 18 different companies and asked, "How many days is your most popular bike trip?" Here are the answers he received:

3, 6, 7, 5, 10, 7, 4, 2, 3, 3, 5, 14, 5, 7, 12, 4, 3, 6

a. Make a line plot of the data.

b. Find the range, median, mean, and mode of the data.

c. On the basis of this information, should Ocean and History Bike Tours change the length of the five-day trip? Explain your reasoning.

7. Which of the following graphs best represents the relationship between a person's age and height? Explain your choice. If you feel that none of the graphs shows this relationship, draw and explain your own graph.

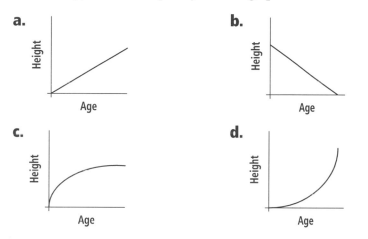

a.
Height / Age

b.
Height / Age

c.
Height / Age

d.
Height / Age

Extensions

8. The number of hours of daylight in a day changes throughout the year. We say that the days are "shorter" in winter and "longer" in summer. The following table shows the number of daylight hours in Chicago, Illinois, on a typical day during each month of the year (January is month 1, and so on).

Month	Daylight hours
1	10.0
2	10.2
3	11.7
4	13.1
5	14.3
6	15.0
7	14.5
8	13.8
9	12.5
10	11.0
11	10.5
12	10.0

a. Describe any relationships you see between the two variables.

b. On a grid, sketch a coordinate graph of the data. Put months on the *x*-axis and daylight hours on the *y*-axis. Do you see any patterns?

c. The seasons in the Southern Hemisphere are the opposite of the seasons in the Northern Hemisphere. When it is summer in North America, it is winter in Australia. Melbourne, Australia, is about the same distance south of the equator as Chicago is north of the equator. Sketch a graph showing the relationship you would expect to find between the month and the hours of daylight in Melbourne.

January in Chicago

January in Melbourne

d. Put the (month, daylight) values from your graph in part c into a table.

Mathematical Reflections

In this investigation, you learned about variables. You made tables and graphs to show how different variables are related. These questions will help you summarize what you have learned:

1 In this investigation, you conducted a jumping jack experiment, collected the data in a table, and made a coordinate graph of the data. Your table and graph showed the relationship between two variables. What were the two variables? How did one variable affect the other?

2 **a.** Name some things in the world around you that vary and that can be counted or measured. Name two variables that you think are related.

b. Explain how you could make a graph to show the relationship between the two related variables from part a. How would you decide which variable should be on the x-axis and which should be on the y-axis?

Think about your answers to these questions, discuss your ideas with other students and your teacher, and then write a summary of your findings in your journal.

Graphing Change

Sidney, Liz, Celia, Malcolm, and Theo found they could comfortably ride from 60 to 90 miles in one day. They used these findings, along with a map and campground information, to plan a five-day tour route. The students wondered how the route would actually work for cyclists. For example, rough winds coming off the ocean or lots of steep hills might make the trip too difficult for some riders.

The friends set off to test the proposed tour route. To make sure the trip would appeal to high school students, Sidney asked her 13-year-old brother, Tony, and her 15-year-old sister, Sarah, to come along. The five college students planned to collect data during certain parts of the trip and use their findings to write detailed reports. They could use their reports to improve their plans and to explain the trip to potential customers.

2.1 Day 1: Philadelphia to Atlantic City

The students began their bike tour near the Liberty Bell and Independence Hall in historic Philadelphia, Pennsylvania. Their goal for the first day was to reach Atlantic City, New Jersey. Sidney, Liz, Sarah, Celia, and Malcolm rode their bicycles. Theo and Tony followed along in a van with the camping gear and repair equipment. Tony recorded the distance reading on the van's trip odometer every half hour from 8:00 A.M. to 4:00 P.M. A map for the entire trip, and Tony's recordings from the first day, are given on the next page.

Time (hours)	Distance (miles)
0.0	0
0.5	9
1.0	19
1.5	26
2.0	28
2.5	38
3.0	47
3.5	47
4.0	47
4.5	54
5.0	59
5.5	67
6.0	73
6.5	78
7.0	80
7.5	86
8.0	89

Problem 2.1

Write a report summarizing the data Tony collected on day 1 of the tour. Describe the distance traveled compared to the time. Look for patterns of change in the data. Be sure to consider the following questions:

- How far did the riders travel in the day? How much time did it take them?

- During which time interval(s) did the riders make the most progress? The least progress?

- Did the riders go further during the first half or the second half of the day's ride?

■ Problem 2.1 Follow-Up

Describe any similarities between the jumping jack data you recorded in Problem 1.1 and the data Tony collected.

2.2 Day 2: Atlantic City to Lewes

On the second day of their bicycle trip, the group left Atlantic City and rode five hours south to Cape May, New Jersey. This time, Sidney and Sarah rode in the van. From Cape May, they took a ferry across the Delaware Bay to Lewes, Delaware. They camped that night in a state park along the ocean. Sarah recorded the following data about the distance traveled until they reached the ferry:

Time (hours)	Distance (miles)
0.0	0
0.5	8
1.0	15
1.5	19
2.0	25
2.5	27
3.0	34
3.5	40
4.0	40
4.5	40
5.0	45

Problem 2.2

A. Make a coordinate graph of the (time, distance) data given in the table.

B. Sidney wants to write a report describing day 2 of the tour. Using information from the table or the graph, what could she write about the day's travel? Be sure to consider the following questions:

- How far did the group travel in the day? How much time did it take them?
- During which time interval(s) did the riders make the most progress? The least progress?
- Did the riders go further in the first half or the second half of the day's ride?

C. By analyzing the table, how can you find the time intervals when the riders made the most progress? The least progress? How can you find these intervals by analyzing the graph?

D. Sidney wants to include either the table or the graph in her report. Which do you think she should include? Why?

Problem 2.2 Follow-Up

1. Look at the second point on your graph as you count from the left. We can describe this point with the *coordinate pair* (0.5, 8). The first number in a **coordinate pair** is the value for the *x*-coordinate, and the second number is the value for the *y*-coordinate. Give the coordinate pair for the third point on your graph. What information does this point give?

2. Connecting the points on a graph sometimes helps you see a pattern more clearly. You can connect the points in situations in which it makes sense to consider what is happening in the intervals *between* the points. The points on the graph of the data for day 2 can be connected because the riders were moving during each half-hour interval, so the distance was changing.

a. Connect the points on your graph with straight line segments.

b. How could you use the line segments to help you estimate the distance traveled after $\frac{3}{4}$ of an hour (0.75 hours)?

3. The straight line segment you drew from (4.5, 40) to (5.0, 45) gives you some idea of how the ride might have gone between the points. It shows you how the ride would have progressed if the riders had traveled at a steady rate for the entire half hour. The actual pace of the group, and of the individual riders, may have varied throughout the half hour. These paths show some possible ways the ride may have progressed:

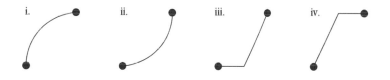

Match each of these connecting paths with the following travel notes.

a. Celia rode slowly at first and gradually increased her speed.

b. Tony and Liz rode very quickly and reached the campsite early.

c. Malcolm had to fix a flat tire, so he started late.

d. Theo started off fast. After a while, he felt tired and slowed down.

2.3 Day 3: Lewes to Chincoteague Island

On day 3 of the tour, the students left Lewes, Delaware, and rode through Ocean City, Maryland, which has been a popular summer resort since the late 1800s. They decided to stop for the day on Chincoteague Island, which is famous for its annual pony auction.

Did you know?

Assateague Island, located next to Chincoteague Island, is home to herds of wild ponies. According to legend, the ancestors of these ponies swam ashore from a Spanish vessel that capsized near the island in the late 1500s. To survive in a harsh environment of beaches, sand dunes, and marshes, these sturdy ponies eat saltmarsh, seaweed, and even poison ivy!

To keep the population of ponies on the island under control, an auction is held every summer. During the famous "Pony Swim," the ponies that will be sold swim across a quarter mile of water to Chincoteague Island.

Celia collected data along the way and used it to make the graph below. Her graph shows the distance the riders were from Lewes as the day progressed. This graph is different from the graph made for Problem 2.2, which represented the *total* distance traveled as the day progressed.

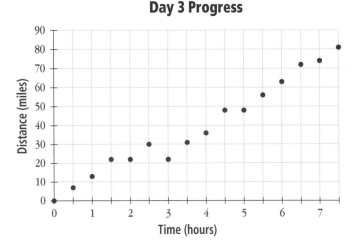

Day 3 Progress

Problem 2.3

A. Would it make sense to connect the points on this graph? Explain.

B. Make a table of (time, distance) data from the information in the graph.

C. What do you think happened between hours 2 and 4? Between hours 1.5 and 2?

D. Which method of displaying the (time, distance) data helps you see the changes better, a table or a graph? Explain your choice.

■ Problem 2.3 Follow-Up

Use the graph to determine the total distance the riders traveled on day 3. Explain how you determined your answer.

2.4 Day 4: Chincoteague Island to Norfolk

On day 4, the group traveled from Chincoteague Island to Norfolk, Virginia. Norfolk is a major base for the United States Navy Atlantic Fleet. Malcolm and Sarah rode in the van. They forgot to record the distance traveled each half hour, but they did write some notes about the trip.

Malcolm and Sarah's Notes

- We started at 8:30 A.M. and rode into a strong wind until our midmorning break.
- About midmorning, the wind shifted to our backs.
- We stopped for lunch at a barbecue stand and rested for about an hour. By this time, we had traveled about halfway to Norfolk.
- At around 2:00 P.M., we stopped for a brief swim in the ocean.
- At around 3:30 P.M., we had reached the north end of the Chesapeake Bay Bridge and Tunnel. We stopped for a few minutes to watch the ships passing by. Since bikes are prohibited on the bridge, the riders put their bikes in the van, and we drove across the bridge.
- We took $7\frac{1}{2}$ hours to complete today's 80-mile trip.

> ## Problem 2.4
>
> **A.** Make a table of (time, distance) data that reasonably fits the information in Malcolm and Sarah's notes.
>
> **B.** Sketch a coordinate graph that shows the same information.

◼ Problem 2.4 Follow-Up

Explain how you used each of the six notes to help you make your table and graph.

2.5 Day 5: Norfolk to Williamsburg

The last stop on the Ocean and History Bike Tour was Williamsburg, Virginia. In America's colonial period, Williamsburg was the capital of Virginia. The buildings of that period have been restored so visitors can imagine what life was like there in the eighteenth century.

After the riders finished lunch, they decided to have a race. The winner would receive $50 from the tour company's first profits. Theo had an electronic speedometer on his bike. It recorded his *speed* every 10 minutes during the 90-minute race.

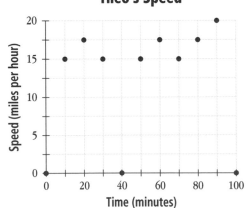

Theo's Speed

Problem 2.5

A. What was Theo's fastest recorded speed, and when did it occur?

B. What was Theo's slowest recorded speed, and when did it occur?

C. Describe the changes in Theo's speed during the race.

D. The graph only shows Theo's speed at 10-minute intervals; it does not tell us what happened between 10-minute marks. The paths below show five possibilities of how Theo's speed may have changed during the first 10 minutes. Explain in writing what each connecting path would tell about Theo's speed.

1.　　　2.　　　3.　　　4.　　　5.

Problem 2.5 Follow-Up

1. Would it be possible for the path below to represent Theo's progress between 10-minute marks? Why or why not?

2. During which 10-minute period(s) of the race did Theo's speed change the most?

As you work on these ACE questions, use your calculator whenever you need it.

Applications

1. Here is a graph of temperature data collected on the students' trip from Atlantic City to Lewes.

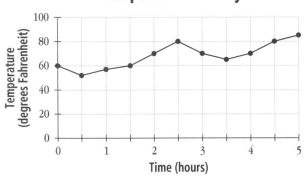

Temperatures for Day 2

a. This graph shows the relationship between two variables. What are they?

b. Make a table of data from this graph.

c. What is the difference between the day's lowest and highest temperature?

d. During which time interval(s) did the temperature rise the fastest? Fall the fastest?

e. Is it is easier to use the table or the graph to answer part c? Why?

f. Is it is easier to use the table or the graph to answer part d? Why?

g. On this graph, what information is given by the lines connecting the points? Is it necessarily accurate information? Explain your reasoning.

2. Katrina's parents kept a record of her growth from her birth until her eighteenth birthday. Their data is shown in the table below.

Age (years)	Height (inches)
birth	20
1	29
2	33.5
3	37
4	39.5
5	42
6	45.5
7	47
8	49
9	52
10	54
11	56.5
12	59
13	61
14	64
15	64
16	64
17	64.5
18	64.5

a. Make a coordinate graph of Katrina's height data.

b. During which time interval(s) did Katrina have her largest "growth spurt"?

c. During which time interval(s) did Katrina's height change the least?

d. Would it make sense to connect the points on the graph? Why or why not?

e. Is it easier to use the table or the graph to answer parts b and c?

3. Make a table and a graph of (time, temperature) data that fit the following information about a day on the road:

   - We started riding at 8 A.M. The day was quite warm, with dark clouds in the sky.
   - About midmorning the temperature dropped quickly to 63°F, and there was a thunderstorm for about an hour.
   - After the storm, the sky cleared and there was a warm breeze.
   - As the day went on, the sun steadily warmed the air. When we reached our campground at 4 P.M. it was 89°F.

4. Amanda is a student at Cartwright Middle School. She is learning how to make graphs. She made the two graphs below to show how her level of hunger and her feelings of happiness changed over the course of a day. She forgot to label the graphs.

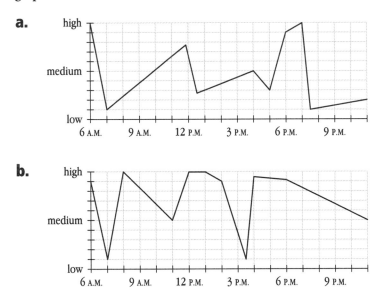

a.

b.

   Here are written descriptions of how the two variables changed throughout the day. Use these descriptions to determine which graph shows the relationship between time and hunger and which graph shows the relationship between time and happiness. Explain your reasoning.

Hunger: Amanda woke up really hungry and ate a large breakfast. She was hungry again by her lunch period, which began at 11:45. After school, she had a snack before basketball practice, but she had a big appetite by the time she got home for dinner. Amanda was full after dinner and did not eat much before she went to bed.

Happiness: Amanda woke up in a good mood, but got mad because her older brother hogged the bathroom. She talked to a boy she likes on the morning bus. Amanda enjoyed her morning classes but started to get bored by lunch time. At lunch, she sat with her friends and had fun. She loves her computer class, which is right after lunch, but then didn't enjoy her other afternoon classes. After school, Amanda had a good time at basketball practice. She spent the evening washing her dog and doing other chores.

5. Here is a graph Celia drew on the bike trip.

   a. What does this graph show?

   b. Is this a reasonable pattern for the speed of a cyclist? Of the van? Of the wind? Explain each of your conclusions.

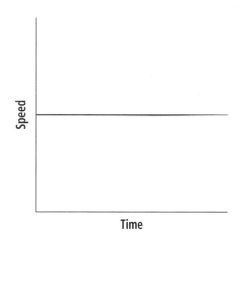

6. The graph below shows the results of a survey of people over age 25 who had completed different levels of education. The graph shows the median salary for people with each level of education.

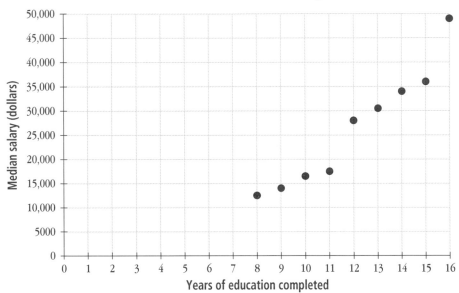

Education and Salary

a. Make a table that shows the information in the graph.

b. After how many years of education do salaries take a big jump? Why do you think this happens?

c. Do you find it easier to answer part b by looking at the graph or your table? Explain your reasoning.

7. When Ben first started to play the electric guitar, his skill increased quite rapidly. Over time, as Ben continued to practice, he seemed to improve more slowly.

a. Sketch a graph to show how Ben's guitar-playing skill progressed over time since he began to play.

b. Your graph shows the relationship between two variables. What are they?

c. What other variables might affect the rate at which Ben's playing improves?

8. Think of something in your life that varies with time, and make a graph to show how it might change as time passes. Some possibilities are the length of your hair, your height, your moods, or your feelings toward your friends.

Connections

9. This table shows the percent of American children in each age group who smoke.

Age	Percent
12	1.7
13	4.9
14	8.9
15	16.3
16	25.2
17	37.2

Source: National Household Survey on Drug Abuse (1991 figures)

a. Make a bar graph of this information.

b. Based on the data, estimate the percent of 18-year-olds who smoke. Explain your reasoning.

c. What relationship does there seem to be between smoking and age? Do you think this pattern continues beyond the teenage years? Explain your reasoning.

10. The following table shows the winners and the winning times for the women's Olympic 400-meter dash since 1964.

Marie-Jose Perec

Year	Name	Time (seconds)
1964	Celia Cuthbert, AUS	52.0
1968	Colette Besson, FRA	52.0
1972	Monika Zehrt, E. GER	51.08
1976	Irena Szewinska, POL	49.29
1980	Marita Koch, E. GER	48.88
1984	Valerie Brisco-Hooks, USA	48.83
1988	Olga Bryzguina, USSR	48.65
1992	Marie-Jose Perec, FRA	48.83

a. Make a coordinate graph of the (year, time) information given in the table. Be sure to choose a scale that allows you to see the differences between the winning times.

b. What patterns do you see in the table and graph? For example, do the winning times seem to be rising or falling? In which year was the best time earned?

Extensions

11. The school booster club sells sweatshirts.

    a. Which, if any, of the following graphs describes the relationship you expect between the price charged for each sweatshirt and the profit made? Explain your choice, or draw a new graph you think better describes this relationship. Explain your reasoning.

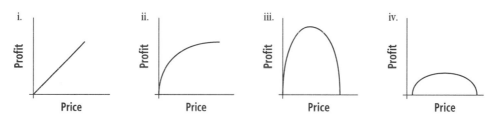

    b. What variables can affect the club's profits?

12. The sketch below shows two bicycles—one with normal wheels, and one with wheels shaped like regular hexagons. Imagine that you put a reflector on the front wheel of each bike and then stood to the side to watch the reflector's path as the bike is ridden past you. Sketch this path:

    a. If the reflector is placed at the center of each front wheel.

    b. If the reflector is placed at the outer edge of each front wheel (near the tire).

13. Chelsea can pedal her bike at a steady pace for long periods of time. Think about how her speed might change if she rode in a wind that fit the pattern shown on the graph below.

a. Sketch a coordinate graph of how Chelsea's speed would change over time if the wind were at her back (a tailwind).

b. Sketch a coordinate graph of how Chelsea's speed would change over time if she had to ride against the wind (a headwind).

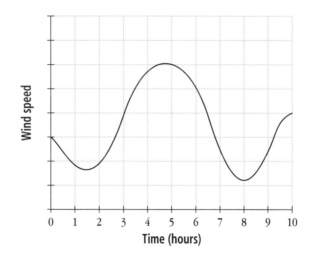

Mathematical Reflections

In this investigation, you analyzed data given in tables, graphs, and written reports. These questions will help you summarize what you have learned:

1 What are the advantages and disadvantages of a table?

2 What are the advantages and disadvantages of a graph?

3 What are the advantages and disadvantages of a written report?

Think about your answers to these questions, discuss your ideas with other students and your teacher, and then write a summary of your findings in your journal.

Analyzing Graphs and Tables

In this investigation, you will continue to use tables and graphs to compare information and make important decisions. Using tables, graphs, and words to represent relationships is an important part of algebra.

Coming up with the idea for Ocean and History Bike Tours was only the first step for the five friends in starting their business. They have other important plans to make as well. Many of these plans involve questions about money.

- What will it cost to operate the tours?
- How much should customers be charged?
- What profit will be left when all the bills are paid?

To answer these questions, the five business partners decided to do some research. They wanted to plan carefully so they would end up with a profit after they had paid all their expenses.

Think about this!

- With your classmates, make a list of the things the tour operators will have to provide for their customers. Estimate the cost of each item per customer.

- How much do you think customers would be willing to pay for the five-day tour?

- Based on your estimates of costs and possible income, will the partners earn a profit?

3.1 Renting Bicycles

The tour operators decided to rent bicycles for their customers rather than having customers bring their own bikes. They called two bike shops and asked for estimates of rental fees.

Rocky's Cycle Center sent a table of weekly rental fees for various numbers of bikes.

Number of bikes	5	10	15	20	25	30	35	40	45	50
Rental fee	$400	535	655	770	875	975	1070	1140	1180	1200

Adrian's Bike Shop sent a graph of their weekly rental fees. Since the rental fee depends on the number of bikes, they put the number of bikes on the x-axis.

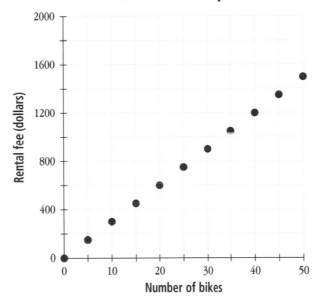

Adrian's Bike Shop Fees

Problem 3.1

A. Which bike shop should Ocean and History Bike Tours use? Explain your choice.

B. Explain how you used the information in the table and the graph to make your decision.

1. In the graph from Adrian's Bike Shop, would it make sense to connect the points with a line? Why or why not?
2. How much do you think each company would charge to rent 32 bikes?
3. Recognizing patterns and using patterns to make predictions are important mathematical skills. Look for patterns in the table and graph on page 37. For each display, describe in words the pattern of change in the data.
4. Based on the patterns you found in part 3, how can you predict values that are not included in the table or graph?

3.2 Finding Customers

Sidney, Liz, Celia, Malcolm, and Theo had a route planned and a bike shop chosen. Now they needed customers. They had to figure out what price to charge so they could attract customers and make a profit.

To help set a price, the partners did some market research. They obtained a list of people who had taken other bicycle tours and asked 100 of them which of the following amounts they would be willing to pay for the Ocean and History Bike Tour: $150, $200, $250, $300, $350, $400, $450, $500, $550, $600. The results are shown in the table.

Tour price	Number who would be customers at this price
$150	76
200	74
250	71
300	65
350	59
400	49
450	38
500	26
550	14
600	0

Problem 3.2

A. If you were to make a graph of the data, which variable would you put on the *x*-axis? Which variable would you put on the *y*-axis? Explain your choices.

B. Make a coordinate graph of the data on grid paper.

C. Based on your graph, what price do you think the tour operators should charge? Explain your reasoning.

■ Problem 3.2 Follow-Up

1. The number of people who said they would take the tour depended on the price. How does the number of potential customers change as the price increases?

2. How is the change in the number of people who said they would go on the tour shown in the table? On the graph?

3.3 Predicting Profit

Based on the results of their survey, the tour operators decided to charge $350 per person for the tour. Of course, not all of this money would be profit. To estimate their profit, they had to consider the expenses involved in running the tour. Sidney estimated these expenses and calculated the expected profit for various numbers of customers. She made the graph below to present her predictions to her partners. Since the profit depends on the number of tour customers, she put the number of customers on the *x*-axis.

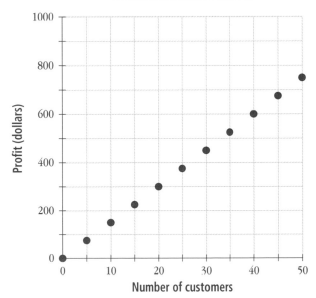

Estimated Tour Profits

Problem 3.3

Study the graph on the previous page.

A. How much profit will be made if 10 customers go on the tour? 25 customers? 40 customers?

B. How many customers are needed for the partners to earn a $200 profit? A $500 profit? A $600 profit?

C. How does the profit change as the number of customers increases? How is this pattern shown in the graph?

D. If the tour operators reduced their expenses but kept the price at $350, how would this change the graph?

■ **Problem 3.3 Follow-Up**

In the profit graph, points at the intersection of two grid lines, such as (20, 300) and (40, 600), are easy to read. Use the "easy to read" points to figure out what the profit would be if only 1 customer went on the tour. How about 2 customers? 3 customers? 100 customers? Describe, in words, the estimated profit for any number of customers.

3.4 Paying Bills and Counting Profits

Sidney was nervous about the partners using her rough estimates to make important decisions. She decided to look more carefully at the company's expected costs and the resulting profit. She found that although the trip would bring in $350 from each rider, it would have operating costs of $30 for each person's bike rental, $125 for each person's food and camp costs, and $700 per tour to rent the van for the trip. Sidney put her estimated cost and income data in a table. Here is the start of her table:

Number of customers	Income	Bike rental	Food and camp costs	Van rental
1	$350	$30	$125	$700
2	700	60	250	700
3				
4				
5				
6				
7				
8				
9				
10				

Problem 3.4

A. Copy Sidney's table. Complete it to give information about income and estimated costs for up to 10 customers.

B. How does the income column change as the number of customers increases? Explain how you can use this relationship to calculate the income for any number of customers.

C. Add and complete a column for "Total cost" (including bike rental, food and camp costs, and van rental) to your table. How does the total cost change as the number of customers increases? Describe how you can calculate the total cost for any number of customers.

D. Add and complete a column for "Profit." What profit would be earned from a trip with 5 customers? 10 customers? 25 customers?

Problem 3.4 Follow-Up

1. What other patterns of change do you see in the table?

2. What is the least number of customers needed for the tour to make a profit?

3. What do you think is the least number of customers needed to make it worthwhile for the students to run the tour? Explain your answer.

As you work on these ACE questions, use your calculator whenever you need it.

Applications

1. This table shows the fees charged for campsites at one of the campgrounds on the Ocean and History Bike Tour:

Number of campsites	1	2	3	4	5	6	7	8
Total campground fee	$12.50	25.00	37.50	50.00	62.50	75.00	87.50	100.00

a. Make a coordinate graph of these data.

b. Would it make sense to connect the points on your graph? Why or why not?

c. Using the table, describe the pattern of change you find in the total campground fee as the number of campsites needed increases. How is this pattern shown in your graph?

2. A camping-supply store rents camping gear for $25 per person.

a. Using increments of 5 campers, make a table showing the total rental charge for 0 to 50 campers. Make a coordinate graph of these data.

b. Compare the pattern of change in your table and graph with patterns you found in the campground fee data in question 1. Describe the similarities and differences between the two sets of data.

3. The tour partners need to rent a truck to transport camping gear, clothes, and bicycle repair equipment. They checked prices at two truck rental companies.

a. East Coast Trucks charges $4.25 for each mile a truck is driven. Make a table of the charges for 0, 25, 50, 75, 100, 125, 150, 175, 200, 225, 250, 275, and 300 miles.

b. Philadelphia Truck Rental charges $200 for a week, or any part of a week, and an additional $2.00 for each mile a truck is driven. Make a table of the charges for renting a truck for five days and driving it 0, 25, 50, 75, 100, 125, 150, 175, 200, 225, 250, 275, and 300 miles.

c. On one coordinate grid, plot the charge plans for both rental companies. Use a different color to mark each company's plan.

d. Based on your work in parts a–c, which truck rental company do you think would be the best deal for the partners?

4. Dezi's family just bought a VCR. Dezi's mom asked him to research rental prices at local video stores. Source Video has a yearly membership package. The manager gave Dezi this table of prices:

Source Video Membership/Rental Packages

Number of videos rented	0	5	10	15	20	25	30
Total cost	$30	35	40	45	50	55	60

Extreme Video does not have a membership package. Dezi made the graph below to show how the cost at Extreme Video is related to the number of videos rented.

Video Rentals from Extreme Video

a. If both video stores have a good selection of movies, and Dezi's family plans to watch about two movies a month, which video store should they choose?

b. Write a paragraph explaining to Dezi how he could decide which video store to use.

c. For each store, describe the pattern of change relating the number of videos rented to the cost.

Connections

5. This summer Jamie is going to Washington, D.C., to march in a parade with his school band. He plans to set aside $25 at the end of each month to use for the trip. Which of the following graphs shows how Jamie's savings will build as time passes?

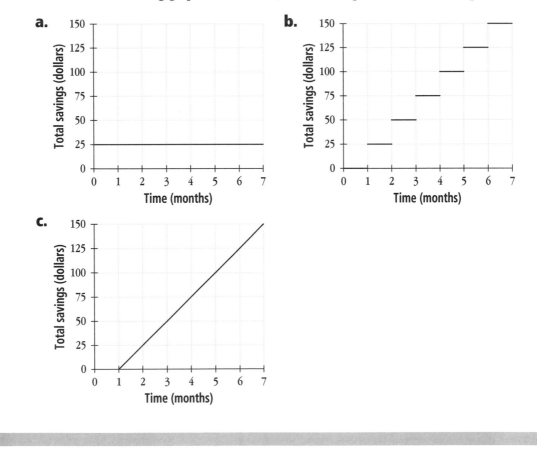

6. Jacy works at a department store on the weekends. This graph represents Jacy's parking expenses. Describe what the graph tells you about the costs for the parking garage Jacy uses.

Parking Costs

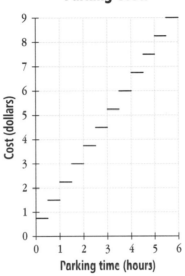

7. Recall that the area of a rectangle is its length times its width.

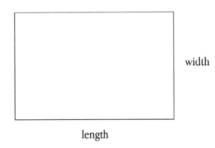

width

length

a. Make a table of the different whole-number values for the length and width of a rectangle with an area of 24 square meters.

b. Make a coordinate graph of your data from part a. Put length on the x-axis and width on the y-axis.

c. Describe what happens to the width as the length increases.

8. Recall that the perimeter of a rectangle is the sum of its side lengths.

    a. Make a table of all the possible whole-number values for the length and width of a rectangle with a perimeter of 24 meters.

    b. Make a coordinate graph of your data from part a. Put length on the *x*-axis and width on the *y*-axis.

    c. Describe what happens to the width as the length increases.

    d. Would it make sense to connect the points in this graph? Explain your reasoning.

9. Here are the box office earnings (in millions of dollars) for a popular movie after each of the first eight weeks following its release.

Weeks in theaters	1	2	3	4	5	6	7	8
Weekly earnings (millions of $)	16	22	18	12	7	4	3	1

    a. Make a coordinate graph showing the weekly earnings after each week. Since a film's weekly earnings depend on the number of weeks it is in theaters, put the weeks in theaters on the *x*-axis and the weekly earnings on the *y*-axis.

    b. Write a short description of the pattern of change in the data table and in your graph. Explain how the movie's weekly earnings changed as time passed, how this change is shown in the table and the graph, and why this change might have occured.

    c. What were the *total earnings* of the movie in the eight weeks?

    d. Make a coordinate graph showing the total earnings after each of the first eight weeks.

    e. Write a short description of the pattern of change in your graph of total earnings. Explain how the movie's total earnings changed over time, how this change is shown in the table and the graph, and why this change might have occurred.

Extensions

10. You can use *Turtle Math* software to draw regular polygons. At each vertex of a polygon, the turtle must make a turn. The size of the turn is related to the number of sides in the polygon. To draw an equilateral triangle, for example, you have to make 120° turns at each vertex.

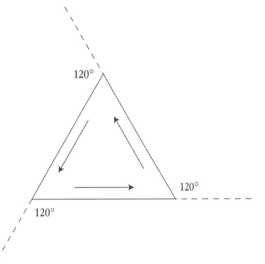

a. Copy and complete the following table, which shows how a turtle turn is related to the number of sides in a regular polygon.

Number of sides	3	4	5	6	7	8	9	10
Degrees in turn	120°							

b. Make a coordinate graph of the (sides, degrees) data.

c. What pattern of change do you see in the degrees the turtle must turn as the number of sides increases? How is that pattern shown in the table? In the graph?

Mathematical Reflections

In this investigation, you learned to use data presented in tables and graphs to help you describe patterns of change in two related variables. You used patterns of change to describe how to predict the value of one variable from the value of the other variable. These questions will help you summarize what you have learned:

1. Imagine a situation in which variable y depends on variable x (for example, y might be profit and x the number of items sold). If y increases as x increases, how would this be indicated in a table? In a graph?

2. If variable y decreases as variable x increases (for example, y might be the amount of money in your wallet on a trip and x the time you have been traveling), how would this be indicated in a table? In a graph?

3. In a coordinate graph of two related variables, when do the points lie in a straight line?

4. In a coordinate graph of two related variables, when is it appropriate to connect the points?

Think about your answers to these questions, discuss your ideas with other students and your teacher, and then write a summary of your findings in your journal.

Patterns and Rules

So far in this unit, you have studied many of the variables involved in the Ocean and History Bike Tours business. By using tables and graphs, you have investigated how these variables are related to one another. For example, you explored how the number of customers is related to profit and how the number of hours of riding is related to the distance covered. As you study how variables are related, you are learning about algebra.

Sometimes the relationship between two variables can be described with a simple **rule**. Such rules are very helpful in making predictions for values that are not included in a table or a graph of a set of data. In previous investigations, you described rules in words. In this investigation, you will use symbols to express rules. Here are some examples:

- If the tour operators charge $350 per customer, the rule for calculating the tour income can be expressed as:

$$\text{income} = 350 \times \text{number of customers}$$

This rule gives the relationship between income and the number of customers: the income is 350 times the number of customers.

- The rule for calculating the circumference of a bicycle wheel (or any circle) can be written as:

$$\text{circumference} = \pi \times \text{diameter}$$

You can use this rule to calculate the circumference of any circle, as long as you know its diameter.

Rules, like those above, that are expressed with mathematical symbols are sometimes referred to as **equations** or **formulas**.

A shorter way to write rules relating variables is to replace the word names for the variables with single letters. For example, in the rule for income, you could write I for *income* and n for the *number of customers.* The rule would then become:

$$I = 350 \times n$$

If you let C stand for *circumference* and d stand for *diameter,* you could write the rule for the circumference of a wheel as:

$$C = \pi \times d$$

You can shorten these rules even more. In algebra, when a number is multiplied by a variable, the multiplication sign is often omitted. So, you could write the above rules as:

$$I = 350n \quad \text{and} \quad C = \pi d$$

When you see a rule, such as $I = 350n$, with a number next to a letter, multiply. So, $I = 350n$ means $I = 350 \times n$ and $C = \pi d$ means $C = \pi \times d.$

 ## 4.1 Heading Home

When the Ocean and History Bike Tour reached Williamsburg, the tired riders packed their bikes and gear in the van and headed back toward Philadelphia. They traveled by interstate highway, and averaged a steady 55 miles per hour for the 310-mile trip home.

You have seen that making a table and a graph can help you understand how the time and the distance traveled are related.

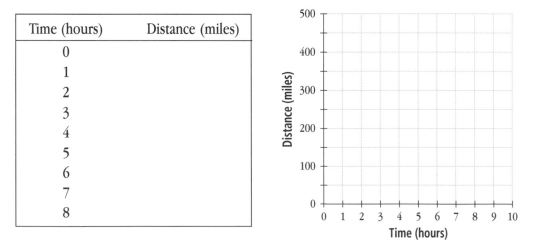

Time (hours)	Distance (miles)
0	
1	
2	
3	
4	
5	
6	
7	
8	

Problem 4.1

A. Copy and complete the table and graph on the previous page to show the relationship between distance and time if the students traveled at a rate of 55 miles per hour.

B. Use your table and graph to estimate the total distance traveled after

    1. 3 hours   **2.** $4\frac{1}{2}$ hours   **3.** $5\frac{1}{4}$ hours

C. If the students continued driving at a steady 55 miles per hour, how far would they go in

    1. 10 hours   **2.** $12\frac{1}{3}$ hours   **3.** 15 hours

D. Look for patterns in the table and graph that help you calculate the distance traveled for any given time. Write a rule, using words, that explains how to calculate the distance traveled for any given time.

E. Use symbols to write your rule from part D as an equation.

Problem 4.1 Follow-Up

1. How could you find the distance traveled after $3\frac{1}{4}$ hours by using the table? The graph? The equation?

2. Estimate how much time it took the students to reach each of the following cities on their route:

    a. Richmond, Virginia—about 55 miles from Williamsburg

    b. Baltimore, Maryland—about $\frac{3}{4}$ of the way from Williamsburg to Philadelphia

    c. Philadelphia, Pennsylvania—about 310 miles from Williamsburg

4.2 Changing Speeds

The speed limit on many sections of the interstate highway is 65 miles per hour. If the students had traveled at this speed for the whole trip, it would have taken them less time to get home. However, if they had stopped for rest and food breaks, they would have probably averaged a slower speed, such as 50 miles per hour.

Problem 4.2

A. Make tables of time and distance data, similar to the table you made for Problem 4.1, for travel at 50 miles per hour and 65 miles per hour.

Plot the data from both tables on one coordinate grid. Use a different color for each set of data. Using a third color, add data points for the times and distances traveled at 55 miles per hour (from Problem 4.1).

B. How are the tables for the three speeds similar? How are they different?

C. How are the graphs for the three speeds similar? How are they different?

D. 1. Look at the table and graph for 65 miles per hour. What pattern of change in the data helps you calculate the distance for any given time? In words, write a rule that explains how to calculate the distance traveled for any given time.

　　2. Use symbols to write your rule as an equation.

E. 1. Now write a rule, in words, that explains how to calculate the distance traveled for any given time when the speed is 50 miles per hour.

　　2. Use symbols to write your rule as an equation.

F. How are the rules for calculating distance for the three speeds similar? How are they different?

■ **Problem 4.2 Follow-Up**

1. After arriving in Philadelphia, Malcolm took the interstate home. He wrote the equation $d = 60t$ to represent his trip home. Explain this equation in words.

2. How long would it take to reach Philadelphia—310 miles from Williamsburg— traveling at 50 miles per hour? 60 miles per hour? 65 miles per hour?

4.3　Calculating Costs and Profits

Sidney started a table like the one on the next page to help the partners determine their profit for the tour. In Problem 3.4, you completed this table for up to 10 customers. You also wrote rules, in words, describing the patterns of change you found in the table.

Sidney wants to use symbols to write equations for these rules, so she can predict costs and profit for any number of customers. For example, in the introduction of this unit the equation for the rule for calculating income was given as $I = 350 \times n$, or $I = 350n$, where I represents the income in dollars for n customers.

Number of customers	Income	Bike rental	Food and camp costs	Van rental	Total cost	Profit
1	$350	$30	$125	$700	$855	⁻$505
2	700	60	250	700	1010	⁻310
3	1050	90	375	700	1165	⁻115
4	1400	120	500	700	1320	80
5	1750	150	625	700	1475	275
6	2100	180	750	700	1630	470
7	2450	210	875	700	1785	665
8	2800	240	1000	700	1940	860
9	3150	270	1125	700	2095	1055
10	3500	300	1250	700	2250	1250

Problem 4.3

A. Write an equation for the rule to calculate each of the following costs for any number, n, of customers.

1. bike rental   **2.** food and camp costs   **3.** van rental

B. Write an equation for the rule to determine the *total cost* for any number, n, of customers.

C. Write an equation for the rule to determine the *profit* for any number, n, of customers.

Problem 4.3 Follow-Up

1. Theo's father has a van he will let the students use at no charge. Which of these equations represents the total cost if they use his van?

   a. $C = 125 + 30$ **b.** $C = 125n + 30n$

   c. $C = 155$ **d.** $C = 155 + n$

2. If the partners require customers to supply their own bikes, which of these is the new equation for total cost? (Assume the students will rent a van.)

   a. $C = 125n + 700$ **b.** $C = 125 + 700 + n$

   c. $825n$ **d.** $C = 350n + 125n + 700$

3. If customers must supply their own bikes, which equations below represent the profit? (Assume the students will rent a van.)

   a. $P = 350 - (125 + 700 + n)$ **b.** $P = 350n - 125n + 700$

   c. $P = 350n - (125n + 700)$ **d.** $P = 350n - 125n - 700$

As you work on these ACE questions, use your calculator whenever you need it.

Applications

1. The El Paso Middle School girls' basketball team drove from El Paso to San Antonio for the Texas State Championship game. The trip was 560 miles. Their bus averaged 60 miles per hour on the trip.

a. Make a table and a graph of time and distance data for the basketball team's bus.

b. Estimate the distance traveled by the bus after each of the following times:

i. 2 hours   **ii.** $2\frac{3}{4}$ hours   **iii.** $3\frac{1}{2}$ hours   **iv.** 7.25 hours

c. How are 2 hours and the distance traveled in 2 hours represented in the table? On the graph?

d. How are $2\frac{3}{4}$ hours and the distance traveled in $2\frac{3}{4}$ hours represented in the table? On the graph?

e. What rule relating time and distance could help you calculate the distance traveled for any given time on this trip?

f. The bus route passed through Sierra Blanca, Texas, which is 90 miles from El Paso. How long did it take the bus to get to Sierra Blanca?

g. The bus route also passed through Balmorhea, Texas, which is $\frac{1}{3}$ of the way from El Paso to San Antonio. How long did it take the bus to get to Balmorhea?

h. How long did it take the bus to complete its 560-mile trip to San Antonio?

2. The equation $d = 70t$ represents the distance, in miles, covered after traveling at 70 miles per hour for t hours.

   a. Make a table that shows the distance traveled, according to this equation, for every half hour between 0 hours and 4 hours.

   b. Sketch a graph that shows the distance traveled between 0 and 4 hours.

   c. If $t = 2.5$ hours, what is d?

   d. If $d = 210$ miles, what is t?

   e. You probably made your graph by plotting points. In this situation, would it make sense to connect these points with line segments?

In 3–6, use symbols to express the rule as an equation. Use single letters to stand for the variables. Identify what each letter represents.

3. The area of a rectangle is its length multiplied by its width.

4. The number of hot dogs needed for the picnic is two for each student.

5. Taxi fare is $2.00 plus $1.10 per mile.

6. The amount of material needed to make the curtains is $4\frac{3}{8}$ yards per window.

7. This table shows the relationship between the number of riders on a bike tour and the daily cost of providing box lunches.

Customers	1	2	3	4	5	6	7	8	9
Lunch cost	$4.25	8.50	12.75	17.00	21.25	25.50	29.75	34.00	38.25

   a. Write an equation for a rule relating lunch cost, L, and number of customers, n.

   b. Use your equation to find the lunch cost if 25 people are on the tour.

   c. How many people could eat lunch if the tour leader had $89.25?

Connections

In 8 and 9, use the following information: In previous units, you discovered that the circumference, radius, and diameter of a circle are related. These relationships involve a special number named with the Greek letter π. The exact value of π is an infinite decimal that begins 3.14159265358. The approximation 3.14 is commonly used. For any circle:

$$\text{circumference} = \pi \times \text{diameter}$$

Since the diameter of a circle is twice its radius, you can also write this as:

$$\text{circumference} = \pi \times 2 \times \text{radius}$$

8. The wheels on Kai's bike are 27 inches in diameter. His little sister, Masako, has a bike with wheels that are 20 inches in diameter. Sometimes Kai and Masako go out for evening bike rides around their neighborhood.

   a. How far will Kai go in one complete turn of his wheels?

   b. How far will Masako go in one complete turn of her wheels?

   c. How far will Kai go in 500 turns of his wheels?

   d. How far will Masako go in 500 turns of her wheels?

e. How many times will Kai's wheels have to turn to cover 100 feet? (Remember that there are 12 inches in 1 foot.)

f. How many times will Masako's wheels have to turn to cover 100 feet?

g. How many times will Kai's wheels have to turn to cover 1 mile? (Remember that there are 5280 feet in 1 mile.)

h. How many times will Masako's wheels have to turn in order to cover 1 mile?

9. The old-fashioned bicycle shown here is called a "penny farthing" bicycle. These bikes had front wheels as large as 5 feet in diameter! Suppose the front wheel of this bicycle has a diameter of 5 feet.

a. What is the radius of the front wheel?

b. How far will this bike travel in 100 turns of the front wheel?

c. How many times will the front wheel turn in a 3-mile trip?

d. Compare the number of times the wheels of Masako's bike turn in a 1-mile trip (see question 8h) with the number of times the front wheel of this penny farthing bike turns in a 3-mile trip. Why do the numbers compare this way?

10. Celia came up with the equation $d = 8t$ to represent the number of miles the bikers could travel in t hours at a speed of 8 miles per hour.

a. Make a table that shows the distance traveled every half hour, up to 5 hours, if the bikers ride at this constant speed for 5 hours.

b. How far would the bikers travel in 1 hour? 3 hours? 4.5 hours? 6 hours?

11. Sean just bought a new CD player and speakers from the Audio Source for $315. The store offered Sean an interest-free payment plan that allows him to pay in weekly installments of $25.

    a. How much will Sean still owe after one payment? After two payments? After three payments?

    b. Using n to stand for the number of payments and A for the amount still owed, write an equation for calculating A for any given value of n.

    c. Use your equation to make a table and a graph showing the relationship between n and A.

    d. As n increases by 1, how does A change? How is this change shown in the table? On the graph?

    e. How many payments will Sean have to make in all? How is this shown in the table? How is this shown on the graph?

Extensions

12. **a.** If you know the distance and the time you have traveled on a car trip, you can calculate the average speed of the trip. Find the average speed for each pair of distance and time values below.

Distance (miles)	Time (hours)	Average speed (miles per hour)
145	2	_____
110	2	_____
165	2.5	_____
300	5.25	_____
446	6.75	_____
528	8	_____
862	9.5	_____
723	10	_____

    b. Write an equation for calculating the average speed, s, for any distance, d, and time, t.

13. The trip from Ocean City, Maryland, to Chincoteague Island, Virginia, is about 40 miles.

    a. How long will the trip take if the riders average 6 miles per hour?

    b. How would the time for the trip change if the average speed increased? If the average speed decreased?

14. Maurice and Cheri made graphs of the equation $y = 4x + 20$ in their math class.

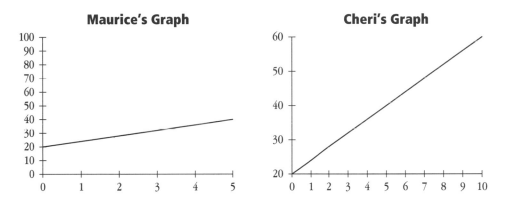

    a. Describe the similarities between the two graphs.

    b. Describe the differences between the two graphs.

    c. Can both of these graphs represent the same equation? Explain your reasoning.

Mathematical Reflections

In this investigation, you learned to use symbols to write equations for rules relating variables. These questions will help you summarize what you have learned:

1. The Larson family is traveling from Michigan to Florida at an average speed of 60 miles per hour. Write an equation for a rule you can you use to calculate the distance they have traveled after any given number of hours.

2. What are the advantages of having an equation to represent the Larson family's situation? What are the advantages to having a table? A graph?

Think about your answers to these questions, discuss your ideas with other students and your teacher, and then write a summary of your findings in your journal.

Using a Graphing Calculator

In the last investigation, you wrote equations to describe patterns and to show how variables are related. Such equations are used in mathematics, science, economics, and many other subject areas. So far, you have written equations that fit the patterns you observed in tables and graphs. Sometimes, you will need to create a table or graph that fits a given equation. In this investigation, you will use a tool called a *graphing calculator* to make graphs and tables that fit a given equation.

5.1 Graphing on a Calculator

The organizers of the Ocean and History Bike Tour need to bring spare parts in the van in case any of the customers have problems with their bikes. They think they will have enough tires if they bring two spare tires for each customer. Theo wrote this rule as the equation $t - 2c$, where t is the number of spare tires needed for c customers.

You can use a graphing calculator to make a graph of Theo's equation. To use a calculator to make graphs and tables, you need to type in an equation that uses the letters y and x to represent the variables. We write these equations so that x is the independent variable and y is the dependent variable. Rewriting Theo's equation using x and y, we get $y = 2x$, where y is the number of tires needed for x customers.

When you press the $\boxed{Y=}$ button, you get a screen like the one below. The calculator gives the "$y =$" part of the equation; you need to type in the rest. You can enter Theo's equation, $y = 2x$, by pressing $\boxed{2}$, followed by the letter $\boxed{X}$.

```
Y₁=2X
Y₂=
Y₃=
Y₄=
```

After you enter the equation, press $\boxed{\text{GRAPH}}$ to see the graph. Here is the graph of $y = 2x$.

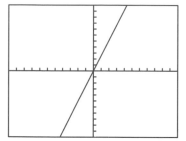

Problem 5.1

Experiment with your graphing calculator and the following equations. Graph one set of equations at a time. For each set, two of the graphs will be similar in some way, and one of the graphs will be different. Answer questions A and B for each set.

Set 1: $y = 3x - 4$ $y = x^2$ $y = 3x + 2$

Set 2: $y = 5$ $y = 3x$ $y = 1x$

Set 3: $y = 2x + 3$ $y = 2x - 5$ $y = 0.5x + 2$

Set 4: $y = 2x$ $y = 2 \div x$ $y = x + 5$

A. 1. Which two equations in the set have graphs that are similar?

   2. In what ways are the two graphs similar?

   3. In what ways are the equations for the two graphs similar?

B. 1. Which equation in the set has a graph that is different from the graphs of the other equations?

   2. In what way is the graph different from the other graphs?

   3. In what way is the equation different from the other equations?

■ Problem 5.1 Follow-Up

1. Use the equation $y = 2x$ to answer the following questions.

   a. If $x = 2$, what is y?

   b. If $x = \frac{2}{3}$, what is y?

   c. If $x = 3.25$, what is y?

   d. You can make a table to show pairs of numbers that fit an equation. Complete the following table for the equation $y = 2x$.

x	0	1	2	3	4	5	6
y							

2. Complete the following table for the equation $y = 2x + 3$.

x	0	1	2	3	4	5	6
y							

3. How is the table for $y = 2x + 3$ in question 2 similar to the table for $y = 2x$ in question 1?

5.2 Making Tables on a Calculator

Some graphing calculators can create tables of data for an equation. To use your calculator to create a table, first press [Y=] and type in an equation. Then, press [TABLE] to see the table for that equation. Here is part of the table for the equation $y = 2x$.

X	Y1
0	0
1	2
2	4
3	6
4	8
5	10
X=0	

Problem 5.2

A. 1. Use your calculator to make a table for the equation $y = 3x$.

2. Copy part of the calculator's table onto your paper.

3. Use your table to find y if $x = 5$.

B. 1. Use your calculator to make a table for the equation $y = 0.5x + 2$.

2. Copy part of the calculator's table onto your paper.

3. Use your table to find y if $x = 5$.

■ Problem 5.2 Follow-Up

1. Use your calculator to make a graph for the equation $y = 3x$. Describe the graph.

2. Use your calculator to make a graph for the equation $y = 0.5x + 2$. Describe the graph.

3. How do the graphs for questions 1 and 2 compare?

4. How would you make a graph for the equations $y = 3x$ and $y = 0.5x + 2$ without a graphing calculator?

As you work on these ACE questions, use your calculator whenever you need it.

Applications

1. Trevor entered an equation into his graphing calculator, and it displayed this table and graph.

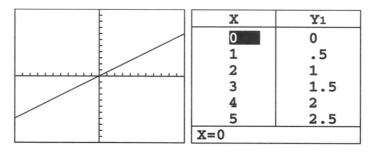

X	Y1
0	0
1	.5
2	1
3	1.5
4	2
5	2.5

X=0

   a. If $x = 5$, what is y?

   b. How is this shown on the table? On the graph?

   c. What equation did Trevor enter into his calculator?

2. Ziamara used her calculator to make a graph of the equation $y = 4x$. She noticed that the point (0, 0) was on the graph. Name three other points that are on the graph. Explain how you found these points.

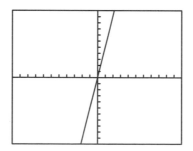

3. Each of the following tables shows how two variables are related. Find a pattern in each table. Use the pattern to complete the missing entries. Express the rule for the pattern as an equation, using the given letters as variable names.

a.

A	0	1	2	3		8	20	100
B	0	7	14	21	28			

b.

X	0	1	2	3	4	8	20	100
Y	6	7	8	9				

c.

X	0	1	2	3	4	8	20	100
Y	1	3	5	7				

d.

R	0	1	2	3	4	6	10	20
S	0	1	4	9	16			

Connections

4. You have seen that many of the costs for the Ocean and History Bike Tour depend on the number of customers. This table shows the relationship between the number of customers and the cost of the ferry ride from Cape May, New Jersey, to Lewes, Delaware.

Customers	1	2	3	4	5	6	7	8	9
Ferry cost	$2.50	5.00	7.50	10.00	12.50	15.00	17.50	20.00	22.50

a. Write an equation relating ferry cost, f, and number of customers, n.

b. Use your equation to find the cost if 35 people are on the tour.

c. How many people could cross on the ferry if the tour leader had $75?

5. Look back at question 4 on page 28 in Investigation 2. The first graph shows the relationship between Amanda's hunger and the time of day. Could you represent this relationship in a table? Could you represent this relationship with an equation? Explain your reasoning.

6. The rules for calculating area and perimeter for common polygons are often written with symbols. Using *A* for area, *P* for perimeter, *b* for base, *h* for height, *l* for length, and *w* for width, write equations for the rules for finding the area and perimeter of each figure below. These equations are usually called *formulas*.

a.

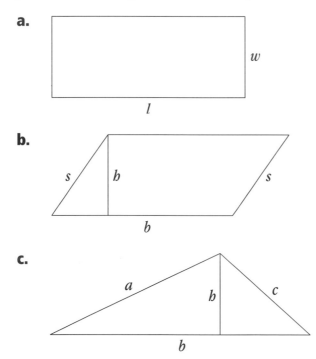

b.

c.

Extensions

7. When the tour partners had a 30-mile race on the last day, they gave the two young riders, Tony and Sarah, a half-hour head start. For this first half hour, Tony and Sarah rode at a steady pace of 12 miles per hour. After their half-hour head start, they kept up a fairly steady pace of about 10 miles per hour. When the others started riding, they went at a fairly steady pace of about 15 miles per hour.

a. Make a table and a graph showing the relationship between distance and time for each group of riders.

b. Will the older riders catch up with Tony and Sarah before the end of the 30-mile race? Explain your answer using both the tables and the graphs.

c. Use d for distance traveled (in miles) and t for riding time (in hours) from when the second group started riding to write equations showing the relationship between these two variables for

i. Tony and Sarah   **ii.** The other riders

Mathematical Reflections

In this investigation, you learned how to use a graphing calculator to make graphs and tables from equations. These questions will help you summarize what you have learned:

1. Write a letter to a friend explaining how to use a graphing calculator to make graphs and tables. Use a specific example to illustrate your explanation.

2. The number of tents the tour organizers need is $\frac{1}{2}$ times the number of customers.

   a. Write an equation for a rule you can use to calculate the number of tents for any number of customers.

   b. Does your equation give you enough information to make a table and a graph? Why or why not?

3. Think of a situation for which you can make a graph and a table, but not an equation.

   Think about your answers to these questions, discuss your ideas with other students and your teacher, and then write a summary of your findings in your journal.

Glossary

change To become different. For example, temperatures rise and fall, prices increase and decrease, and so on. In mathematics, quantities that change are called *variables*.

coordinate graph A graphical representation of pairs of related numerical values that shows the relationship between two variables. It relates the independent variable (shown on the *x*-axis) and the dependent variable (shown on the *y*-axis).

coordinate pair An ordered pair of numbers used to locate a point on a coordinate grid. The first number in a coordinate pair is the value for the *x*-coordinate, and the second number is the value for the *y*-coordinate.

dependent variable One of the two variables in a relationship. Its value depends upon or is determined by the other variable called the independent variable. For example, the cost of a long distance phone call (dependent variable) depends on how long you talk (independent variable).

equation, formula A rule containing variables that represents a mathematical relationship. An example is the formula for finding the area of a circle: $A = \pi r^2$.

independent variable One of the two variables in a relationship. Its value determines the value of the other variable called the dependent variable. If you organize a bike tour, for example, the number of people who register to go (independent variable) determines the cost for renting bikes (dependent variable).

pattern A change that occurs in an predictable way. For example, the squares on a checkerboard form a pattern in which the colors of the squares alternate between red and black. The sequence of square numbers: 1, 4, 9, 16, . . . forms a pattern in which the numbers increase by the next odd number. That is, 4 is 3 more than 1, 9 is 5 more than 4, 16 is 7 more than 9, and so on.

relationship An association between two or more variables. If one of the variables changes, the other variable may also change, and the change may be predictable.

rule A summary of a predictable relationship that tells how to find the value of a variable. It is a pattern which is consistent enough to be written down, made into an equation, graphed, or made into a table. For example, this rule relates time, rate, and distance: distance is equal to rate times time, or $d = rt$.

scale A labeling scheme used on each of the axes on a coordinate grid.

symbolic form Anything written or expressed through the use of symbols. For example, letters and numbers are often used instead of words to represent mathematical rules.

table A list of values for two or more variables that shows the relationship between them. Tables often represent data made from observations, from experiments, or from a series of arithmetic operations. A table may show a pattern of change between two variables that can be used to predict values for other entries in the table.

variable A quantity that can change. Letters are often used as symbols to represent variables in rules or equations that describe patterns.

x-axis The number line that is horizontal on a coordinate grid.

y-axis The number line that is vertical on a coordinate grid.

Index